W9-DDG-077

CAVES

Dedicated to Jon, Sophie, and Ruby — NCB

Dedicated to the two most important women in my life — my wife, Jaclyn, and my mother, JoAnn — and, of course, my dog, Toby — KC

Acknowledgments: Thank you to Devra Heyer, Amanda Willis, Nathan Roser, and Meredith Hall Weberg from the National Speleological Society for fact-checking assistance; "Underground Astronauts" Marina Elliott and K. Lindsay Hunter for sharing their personal experiences discovering *Homo naledi* in the Rising Star cave system; and, always, thank you to my father, Dr. Nicholas Cross, for providing substantial research for this book and raising me to share his love of the natural world. — NCB

Library of Congress Cataloging-in-Publication Data Available

ISBN 978-1-338-72662-6

10 9 8 7 6 5 4 3 2 1 22 23 24 25 26

Printed in China 38
First edition, October 2022

Book design by Sarah Dvojack
Art direction by Brian LaRossa

The text type was set in Americana and Futura. The display type was set in Americana.

CAVES

Written by
Nell Cross Beckerman

Illustrated by
Kalen Chock

Orchard Books ◆ New York

In the shade

of the woods

is a hill

with a hole.

Beaconing black.

Goose-bump chills.

Excitement and fear battle.

Which will win?

You want to
go
in . . .

Do you dare?

A cave is a natural hole big enough for a human to enter. Caves are usually deeper than they are tall. Speleology is the science of studying and exploring caves. Cave explorers are known as cavers or spelunkers. There are amazing caves all over the world waiting to be discovered — just don't go in alone!

Seeping through the earth —

water.

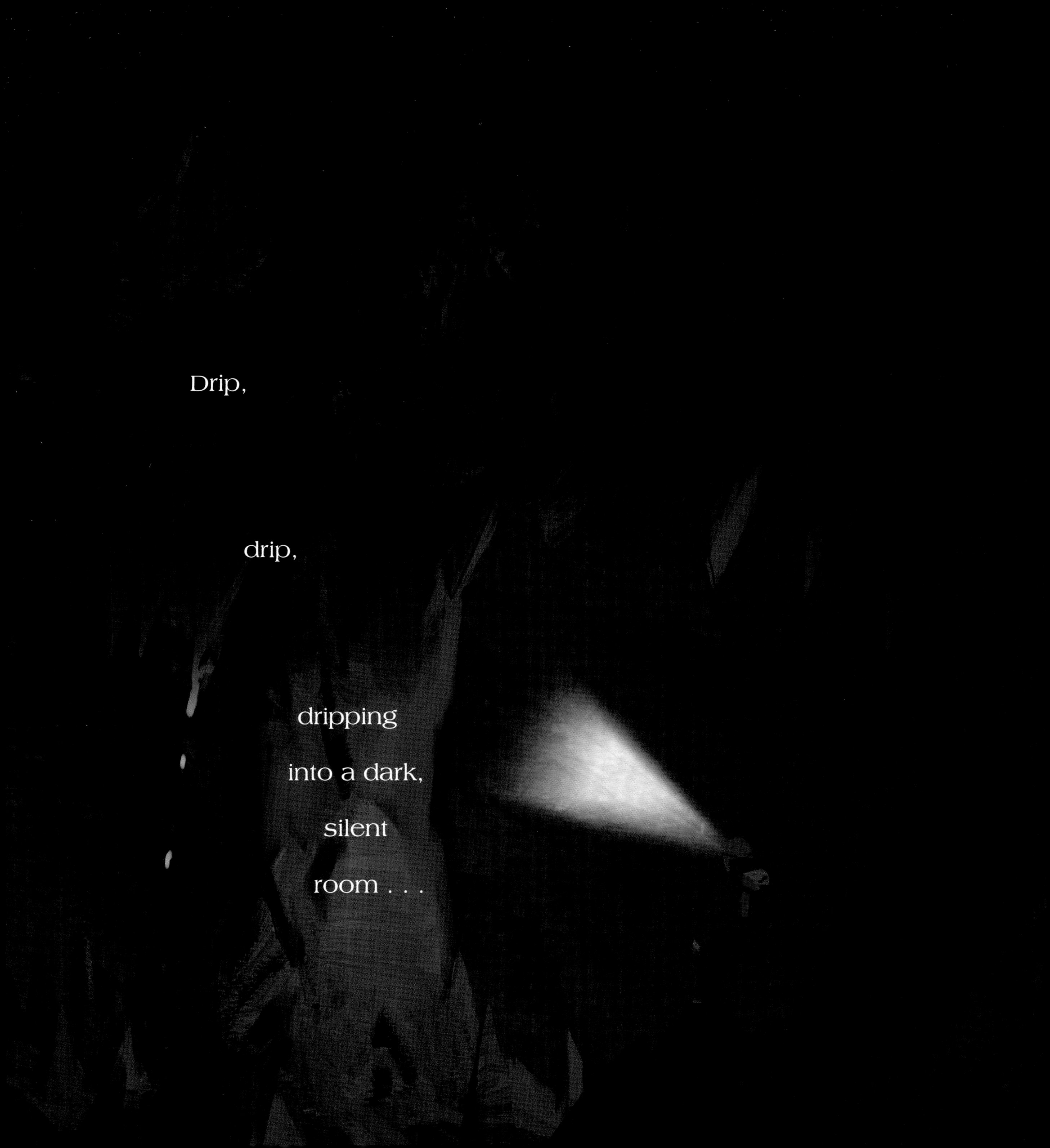

Drip,

drip,

dripping

into a dark,

silent

room . . .

Careful!

When raindrops fall on limestone over millions of years, caves and beautiful rock formations are made. Water dripping underground leaves particles behind on the cave roof, creating rock pendants called stalactites. Directly below the stalactites, the water falls to the ground, building up stone stalagmites. If broken, they could take thousands of years to grow back, if they grow back at all.

This cave sat in secret,

appearing ordinary

except to those few

who

looked closer.

Took time to go

deeper . . .

A dazzling surprise!

In 2000, in Naica, Mexico, two mining brothers discovered caves filled with some of the largest natural crystals ever found.

The Cueva de los Cristales had been filled with water, but after a silver-mining company pumped the water out, people could see the thirty-nine-foot selenite crystals.

Scorching temperatures required visitors to wear special cooling suits to study the crystals. When the pumping stopped, the cave reflooded, refilling with the warm, mineral-rich water that helps the crystals grow.

Six women

dropped into

darkness.

Squeezing,

crawling,

superhero scooching,

to dig up

bones.

Hello, ancient cousin!

More than 1,800 fossils of a previously unknown early relation to humans, *Homo naledi*, were discovered in the Rising Star cave system in South Africa. They were more than 200,000 years old! Studies of these ancient "cousins" continue today and help inform our understanding of human history. Hired for their dexterity, ability to fit in tight spaces, and scientific training, six women carried out the main excavation. One section of the entrance required crawling in a "Superman" pose, with one arm out and the other tight against the chest.

In the humid, hot swamp,

gators swim,

turtles plop,

snakes skim,

water rushes.

Divers sink into the black,

while you . . .

CANNONBALL!

Underwater cave systems are caves connected by long passages. Divers in Florida swam the passages of one cave system for twenty-nine hours — the longest cave dive ever! They entered at the Wakulla Springs Cave and came out seven miles later at the Turner Sink Cave. They mapped what is now called the Wakulla–Leon Sinks cave system. It is the longest underwater cave system in the United States.

When sun sets,

they awake by the thousands,

flying jaggedly

out

of

your

dreams.

Guided by sound to eat,

then returning to their

inverted

slumber.

Stop! Do not enter!

Every summer, twenty million bats roost in Bracken Cave outside of San Antonio, Texas, making it one of the largest known concentrations of mammals in the world.

While bats are often feared or viewed as pests, the majority of bats are harmless and help us by eating insects like mosquitos. Bat conservationists work to protect bats from people by building gates at some cave entrances, letting bats enter and exit but keeping people out.

Ancient cave people,

looked . . . how?

Did . . . what?

So much still a mystery,

but we

keep

finding

clues.

More than twenty thousand years ago, some early humans lived in caves and left behind cave paintings. Several of the most famous are found in the Lascaux Cave in France, which has more than two thousand figures.

Cave dwellers often used charcoal for the color black, and they ground up other minerals and mixed them with water, animal fat, and other substances to make different colors of paint. Early cave painters did not sign their paintings, because writing was not yet invented. Sometimes, they used a tube to blow paint powder over a hand pressed to the wall — these ancient handprints can still be seen today.

Above a river of eels,

a silently squirming ceiling

illuminates the

thirty-million-year-old walls,

dripping an

eerie constellation

to trap snacks.

Wanna share?

The Waitomo Caves in New Zealand are famous for their Glowworm Grotto, where tourists can ride a boat on an underground river through a chamber where the only light comes from bioluminescent glowworms. The worms spin silk on the cave ceiling and drop down as many as thirty silk threads peppered with sticky droplets. The glow attracts insects, which then get stuck, and the worm retracts the thread and then eats the prey alive.

Molten hot earth

spilling out,

burning through

everything.

Leaving

empty

tubes.

Most caves are made slowly, but lava tube caves form quickly! Hot lava cools when it contacts air, forming a shell and allowing the hot lava to flow even faster. Once the flow ends, the tube is empty and cools to volcanic rock. Undara Lava Tubes in Australia formed almost 200,000 years ago, and some are wheelchair accessible.

All over the world,

caves wait.

Filled with our past,

our future.

Undiscovered.

Ready for

wondering,

wandering

explorers

like you

to study

and crouch

and crawl

to find them.

Do you dare?

AUTHOR'S NOTE

I have loved caves ever since I was a kid. My dad often took me on nature adventures, including caving. The caves were cold! Dark! Moist! And I loved the otherworldly rock formations. Caving was so exciting!

As I got older, I went on even more cave adventures. Some of my favorites were exploring the Hana Lava Tube on the Hawaiian island of Maui; visiting sea caves in Capri, Italy; floating in an inner tube through Jaguar Paw cave in Belize; and admiring the glittering limestone formations in Bermuda's Crystal Cave. One summer, I took my family on a road trip where we descended five narrow flights of stairs to see Black Chasm Cavern in California and swam in Coyote Creek Cave.

To write this book, I closed my eyes and thought about how exploring those caves felt. I wanted to use words to re-create those feelings. I also did a lot of research, using books and websites to travel to faraway caves in my imagination.

I hope this book inspires you to see if there is a cave close to where you live, and to go on a nature adventure to see it!

Nell and her dad, Anza-Borrego Desert State Park, 1976

Nell in the Hana Lava Tube, Maui, Hawaii, 2003

ILLUSTRATOR'S NOTE

As a kid, I always felt in tune with the world around me. Whether it was driving through the beautiful hills of Northern California, driving up the coast on Oahu, or visiting old castles in Europe, life has always provided me with great experiences and inspiration for my paintings. Every person and place that I have met and seen on my art journey has had an impact on how I see the world.

Nature has always been an escape for me, and I am very humbled to have illustrated this book, and I hope that it will encourage you to be curious of the world around you. Adventure awaits!

Live Aloha.

Cave Rules

- Never go caving alone. Make sure there is an adult with you.
- Tell someone where you are going and when you will be back.
- Double-check your supplies and dress appropriately.
- If you find a new cave, report it. Don't explore an undiscovered cave without an expert.
- Leave only light footprints and take only pictures — respect the caves and don't damage them. If you find cave paintings or ancient artifacts, report it to a local university, which can learn from them.
- Cave ecosystems are fragile. Take extra care to leave them just as you found them.

Consider joining a local caving group to explore caves in your area. Find out more at caves.org, home of the National Speleological Society.

Spelunking Equipment

- Helmet with light attached
- Ropes
- Compass
- Gloves
- Two flashlights
- Sturdy boots
- First aid kit
- Water and snacks

More Fun Facts About Caves!

GLOWWORM GROTTO, WAITOMO CAVES,
New Zealand

- These caves are thirty million years old!
- They were first explored in 1887 by local Maori chief Tane Tinorau, who was accompanied by English surveyor Fred Mace. Many staff employed at the caves today are direct descendants of Chief Tinorau and his wife, Huti.
- Glowworms have a special characteristic called bioluminescence. This is a chemical reaction in their bodies that allows them to produce light!
- The Waitomo Caves are safe and open for public tours.

CUEVA DE LOS CRISTALES,
Naica, Mexico

- Besides Cueva de los Cristales, there were other important caves discovered in the Naica mine. In 1910, miners discovered the "Cave of Swords," which was filled with crystal daggers coming from the walls. Each was about three feet long!
- When the Cueva de los Cristales was available for exploration, the temperature was almost 120 degrees Fahrenheit (49 degrees Celsius) and the humidity was between 90–100%. To enter the cave in these extreme temperatures, scientists had to invent special suits and breathing systems to allow safe working conditions.

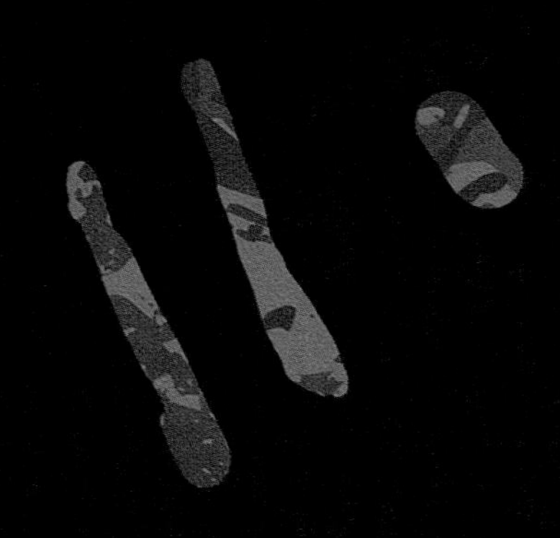

RISING STAR CAVE SYSTEM,
South Africa

- When paleoanthropologist Lee Berger discovered he was too big to access the main excavation site in the Rising Star cave system, he recruited six women, called the "Underground Astronauts," to carry out the excavations. Lee and a research team watched from above on connected TV monitors.
- The Underground Astronauts had to be trained scientists; have experience rock climbing and caving; be unafraid of small, dark spaces; be strong and flexible; and know how to handle fossils. Becca Peixotto, Alia Gurtov, Elen Feuerriegel, Marina Elliott, K. Lindsay Hunter, and Hannah Morris were selected. Shifts in the cave lasted from two to six hours.
- Researchers don't know yet why so many *Homo naledi* skeletons ended up in such a difficult-to-reach cave! They believe it might have been a type of burial ground. Not all researchers agree, but to date, it's the best explanation there is for the evidence from this fascinating site!